Flowering Plants

A Success Story

Rebecca L. Johnson

Sally Ride, Ph.D., President and Chief Executive Officer;
Tam O'Shaughnessy, Chief Operating Officer and
Executive Vice President; Margaret King, Editor;
Monnee Tong, Design and Picture Editor; Erin Hunter,
Science Illustrator; Brenda Wilson, Editorial Consultant;
Matt McArdle, Editorial Researcher

Program Developer, Kate Boehm Jerome
Program Design, Steve Curtis Design Inc.
www.SCDchicago.com

Sally Ride Science
9191 Towne Centre Drive
Suite L101
San Diego, CA 92122

ISBN: 978-1-933798-67-7

Printed in the United States of America
10 9 8 7 6 5 4 3 2 1
First Edition

Cover: Whether simple or complex, flowers are the key to
flowering plants' success.

Title page: A northern red-back vole snacks on sweet wild
blueberry fruits.

Right: A queen butterfly rests on the large drooping
blossom of an angel trumpet plant.

Sally Ride Science is committed to minimizing its environmental impact by using
ecologically sound practices. Let's all do our part to create a healthier planet.

This book is printed on paper made with 100% recycled fiber, 50% post-consumer
waste, bleached without chlorine, and manufactured using 100% renewable energy.

Contents

In Your World

Wow! Look at all those pumpkins! A few months ago, each one was just a big yellow flower. Now they're giant orange fruits jam-packed with seeds.

A pumpkin is a fruit? Technically, yes. Fruits are seed-bearing structures made by **flowering plants**. Apples, beans, walnuts, tomatoes, figs . . . yes, they're all fruits of flowering plants. Earth is home to roughly 400,000 different kinds of them. In fact, flowering plants make up about 91 percent of all the plants on the planet!

What's the secret to their success? Flowers, of course! Flowers give seeds the best possible chance to start the next generation. So how did flowering plants get *their* start? Well, it all began a long, long time ago . . .

How Flowering Plants Conquered the Land

Ever wished you could travel through time? Here's your chance! Imagine a time machine is taking you back, back, back. When it stops, you see Earth as it was 500 million years ago.

The land is rocky and bare. There's no sign of life. But look into the water lapping the shore. It's teeming with tiny green dots. The dots are **algae**. Zillions of these one-celled life forms swarm in Earth's ancient oceans.

Magnification: 150x

▲ **Ancient algae similar to these green algae were among the first living things that could make their own food.**

These algae have a special talent. They can make their own food. Using energy from sunlight, they turn carbon dioxide—a gas—and water into sugar. They also produce oxygen, another gas. The whole process is called **photosynthesis**.

Powered by the food they make, the algae grow and **reproduce**. Some simply split in half to make more of their kind. Others release tiny reproductive structures called **spores** that grow into new algae cells.

Out of the Water

Now you're traveling forward in time. You stop at 475 million years ago. The land along the shore is covered with green stuff. Meet some of the first land plants! Small and low-growing, they hug the ground.

These plants **evolved** from algae. Over millions of years, they changed in ways that let them move onto land. It was a good move. Conditions are better for photosynthesis on land. There's more sunlight than in water. There's more carbon dioxide in air than in water, too.

But these ancient plants can survive only in moist places. They will die if they dry out. The plants need water to reproduce, too. Water carries their spores from place to place. And those spores need water to grow into new plants.

So THAT's Why!

Life appeared in Earth's oceans about 3.7 billion years ago. Why did it take so long for living things to come onto land? For billions of years, deadly **ultraviolet light** from the Sun would have killed any organisms on land. Over time, oxygen released by photosynthesizing algae built up in the air and made it breathable. The oxygen also formed an **ozone layer** high up in the air. The ozone layer blocked most of the ultraviolet rays, and life headed for shore.

▶ The first land plants were simple and small, like these modern-day liverworts.

The Bottom Line The first land plants evolved from green algae that carried out photosynthesis in the oceans.

Plants with Parts

Your time machine shudders to another stop. Where are you now? It's 410 million years ago. What's changed? A lot!

Small branching plants rise up from the ground. Look closely. These plants have tiny leaves wrapped around stems. A network of roots anchors the plants in the soil.

Inside, tiny tubes, or **vessels**, run from roots to stems to leaves. Food made by photosynthesizing leaves travels through these vessels. So does water soaked up by roots. Like pipes, vessels transport food and water to all the plant's cells.

Did you notice that these plants live in slightly drier places than those you saw before? They've got a few tricks up their sleeves . . . or rather, leaves. A waxy coating helps seal in water. Tiny openings let carbon dioxide and oxygen move in and out so photosynthesis can take place. But the openings can close to keep the plants from losing water and drying out.

▲ Leaves were small—really small—when they first appeared in the plant world.

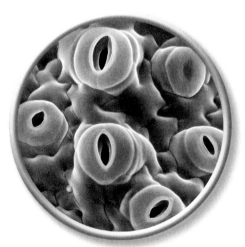

Magnification: 500x

▲ Tiny openings on leaf surfaces allow gases to move in and out of leaves.

First Forests

When your time machine stops again, it's 400 million years ago. Waist-high plants are everywhere. Many look a lot like modern-day ferns—we know that from their **fossils**. They cover the land wherever it's moist enough for them to grow. They reproduce with spores, just like their ancestors did.

Zoom forward another few million years. Many of the plants have gotten huge! Giant horsetails tower overhead. Some stand 30 meters (98.4 feet) tall. Swirls of tiny leaves or short, needle-like branches go around and around their stems.

And how about those ferns! Many types are as tall as trees. Together, these tree ferns and horsetails form shadowy, misty forests—the world's first.

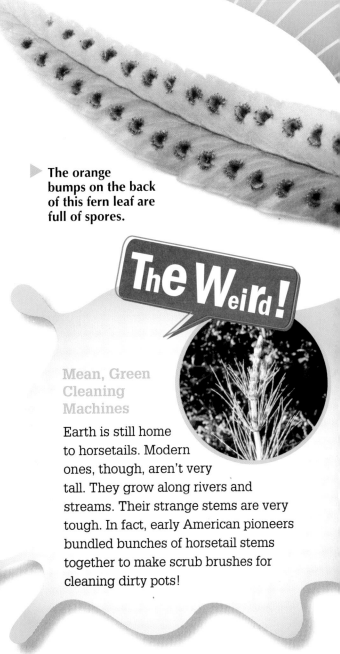

▶ The orange bumps on the back of this fern leaf are full of spores.

The Weird!

Mean, Green Cleaning Machines

Earth is still home to horsetails. Modern ones, though, aren't very tall. They grow along rivers and streams. Their strange stems are very tough. In fact, early American pioneers bundled bunches of horsetail stems together to make scrub brushes for cleaning dirty pots!

The Bottom Line | The development of transport tubes and water-saving features allowed some ancient plants to grow larger and live farther from water.

The Seed Story

Your time machine zooms to 385 million years ago. You step out into another forest. Right away, you spot a plant that is new on the scene. It looks a bit like a fern. But it has something that ferns don't. This plant has **seeds**.

Seeds are a whole new way of reproducing in the plant world. A seed contains an embryo, a miniature version of a plant, all ready to grow. The embryo is surrounded by a starchy food. It's like an energy bar the embryo can snack on when it begins to grow. The embryo and its food are surrounded by a tough **seed coat**. Like a suit of armor, the seed coat protects the delicate embryo inside.

An artist's drawing shows what the oldest known seed plant, *Elkinsia polymorpha*, might have looked like.

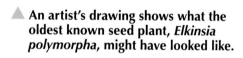

This is a drawing of a bean plant like the kind grown in backyard gardens today. Can you see the tiny plant parts of the embryo inside this seed?

From Egg to Embryo

So how do seeds form? It all starts with tiny **ovules**, which contain even tinier egg cells. These eggs are the female part of what will become seeds.

Next, you need dust-like particles called **pollen**. Pollen contains **sperm**, which are the male part of what will become seeds. Some plants produce only ovules. Some produce only pollen. Others produce both.

When a particle of pollen and an ovule come together, a seed begins to form. A sperm from the pollen joins with an egg in the ovule. This joining is called **fertilization**. Each fertilized egg forms an embryo. What's left of the ovule forms the rest of the seed.

Seeds were a huge step forward for plants on ancient Earth. It wasn't long before seed-making plants began turning up all over the planet.

▲ Small male cones on a pine tree release clouds of dust-like pollen grains.

▶ A female pinecone has ovules on its woody scales. When pollen from a male cone reaches the ovules, seeds form.

The Bottom Line | Plants that produce seeds—tough protective structures with tiny plant embryos inside—evolved about 385 million years ago.

11

Modern-day pine trees evolved from cone-bearing seed plants that lived millions of years ago.

The Cone-Bearers

Remember, spore-making plants can't reproduce without water. But seed plants can. They can make seeds whether their surroundings are wet or dry.

And seeds are tough customers. Their seed coats protect them from heat and cold and from drying out. This means an embryo can wait a long time before starting to grow. It can wait until conditions in the environment around it are just right for growth.

Because seed plants didn't need water to reproduce, they spread into all sorts of places where spore-makers couldn't live. The earliest types of seed plants produced seeds out in the open, on branches or leaves. About 300 million years ago, other plants evolved that produced seeds partly sheltered by **cones**. Cone-making seed plants went on to rule the plant kingdom for many millions of years.

Flower Power

Hang on! You're speeding forward again, this time to 140 million years ago. Cone-bearing plants are still going strong. But there is also something growing here that's completely new. Say "howdy" to the first flowering plants.

Flowers are what make flowering plants, like this bird of paradise, unique.

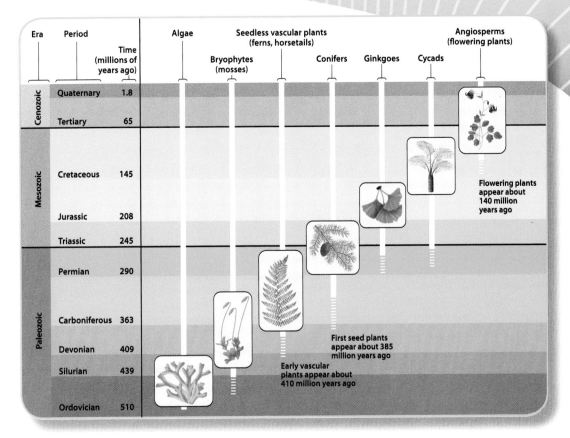

Era	Period	Time (millions of years ago)
Cenozoic	Quaternary	1.8
Cenozoic	Tertiary	65
Mesozoic	Cretaceous	145
Mesozoic	Jurassic	208
Mesozoic	Triassic	245
Paleozoic	Permian	290
Paleozoic	Carboniferous	363
Paleozoic	Devonian	409
Paleozoic	Silurian	439
Paleozoic	Ordovician	510

Column labels across the chart: Algae, Bryophytes (mosses), Seedless vascular plants (ferns, horsetails), Conifers, Ginkgoes, Cycads, Angiosperms (flowering plants)

Flowering plants appear about 140 million years ago

First seed plants appear about 385 million years ago

Early vascular plants appear about 410 million years ago

▲ This chart shows when different kinds of algae and plants got their start. Flowering plants are the latecomers in Earth's history. But today they are the dominant plants on land.

These new arrivals have parts never seen before in the plant world. You guessed it—flowers! Flowers are the secret to how flowering plants reproduce. Seeds form inside a flower. Then the part of the flower that holds the seeds turns into a fruit. Fruits might just be nature's best invention for packaging seeds that the plant world has ever seen! Zoom ahead to 50 million years ago. Flowering plants are everywhere! They have become the most successful plants on Earth. And guess what? Today, they still are!

The Bottom Line

Flowering plants replaced cone-making plants as the most successful plants on Earth.

Meet the Flowering Plants

▲ It's hard to believe the places some flowering plants call home.

Where can you go to get away from flowering plants? Try the North Pole. Or try the deep ocean . . . or the top of Mount Everest. That's right—flowering plants are pretty much everywhere!

Except for the always-frozen places, flowering plants grow in every **habitat** on land—tall trees in the rainforest, waist-high grasses on the prairies, and spiny cactuses in the desert. Some flowering plants live in lakes, streams, and other water habitats, too.

Flowering plants come in all sizes. Duckweed is the smallest. Each floating plant is barely 1 millimeter (0.04 inches) across. The largest flowering plant is a type of eucalyptus tree. The tallest one alive stands 98 meters (322 feet) tall.

Amazing Variety

Got a favorite flower? Roses? Daisies? Maybe you're a tulip fan. With so many flowers to choose from, picking just one is tough.

Flowers come in a rainbow of colors. They come in many sizes, too. Grass flowers are so small you need a magnifying glass to see them. No one, however, could miss the flowers of *Rafflesia arnoldii*. Measuring nearly 1 meter (3.3 feet) across, they are the largest flowers on Earth.

Every kind of flower is different. But a typical flower has colorful petals that surround its reproductive parts. Those parts include a carpel that holds one or more ovules. Around the carpel are skinny stamens that have pollen on their tips.

The Weird!

A Big Stink

So, what does the world's biggest blossom smell like? Careful— don't be too quick to take a sniff. The flowers of *Rafflesia arnoldii* smell like . . . rotten meat. *Eew!*

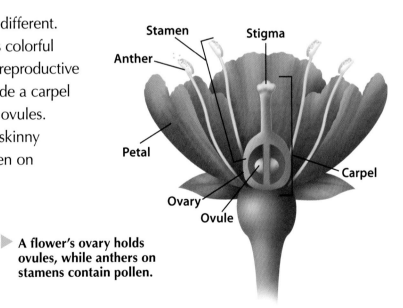

Stamen Stigma

Anther

Petal

Ovary

Ovule

Carpel

▶ A flower's ovary holds ovules, while anthers on stamens contain pollen.

Found in nearly every habitat, flowering plants produce flowers that vary greatly in size, shape, and color, but have similar basic parts.

The Birds and the Bees

Remember, an egg and sperm must meet in order for seeds to form. Most flowering plants need help to make this happen. That's where **pollinators** enter into the picture. They move pollen from one flower to another. When a sperm from the pollen of one type of flower and an egg from the ovule of the same kind of flower come together, seeds form—with new little plants inside.

The major pollinators of flowering plants are insects, birds, and bats. Flowering plants have evolved all sorts of **adaptations** to attract these animal helpers. One adaptation is to produce a sweet treat of **nectar**. The sugary liquid is usually found deep inside a flower. When a pollinator crawls or reaches inside the flower to get the nectar, pollen sticks to its body. When the pollinator moves to another flower of the same type, some of the pollen gets deposited on that flower's sticky stigma. The process of making a seed has begun!

Why does *Rafflesia arnoldii* smell like rotten meat? It's because a fly—the bluebottle carrion fly—is the pollinator of this stinky flower. And this fly thinks stinky is great!

▼ **A bat gets covered with pollen as it moves from flower to flower.**

Yellow "stripes" on these iris petals point the way to nectar deep inside.

Tricks of the Trade

Bright colors are another adaptation that helps attract pollinators to flowers. Flowers with red, yellow, orange, or purple petals are easy to spot from a distance. Pollinators head for them like moths to a flame.

Many flowers also have contrasting stripes or spots on their petals. These designs aren't just for show. They are adaptations to help direct pollinators to the nectar the flowers contain. For a bee in flight, for example, these markings are like the lines on airport runways. They show the bee where to land and exactly where to go.

Some flowers play tricks, though. They have wonderful scents that hint at a sweet prize hidden inside. But they don't produce nectar at all! When pollinators land on these flowers, they crawl around in search of the sugary liquid they think is there. In the process, they get covered with pollen that they'll spread to the next flower, and the next, and the next.

The Bottom Line | Flowering plants have many adaptations to attract pollinators, which help the plants reproduce by spreading their pollen.

Away We Go

Have you ever watched seeds whirling down from a maple tree? Flowers aren't good just at making seeds. They've also evolved remarkable ways of spreading their seeds around.

Spreading seeds means new plants will grow far from their parent plant. That's a good thing. If too many plants grow in the same place, there aren't enough water and **nutrients** to go around.

The wing-like blades of maple seeds make the seeds spin like little helicopters. Spinning slows their fall. Slow-falling seeds have a better chance of being carried away by the wind.

Some seeds travel by water. Coconuts are big seeds covered with a thick but light husk, or covering. Coconuts can float. Those that fall into the ocean are often carried by currents to distant beaches. Some coconuts have made it halfway around the world.

A few plants take a different approach. When their fruits are ripe, they explode. Common garden flowers called impatiens have seed pods that explode when touched. Seeds go whizzing off in every direction and may land meters from the plant.

▼ **Seeds with fluffy "parachute" tops swirl away in the slightest breeze.**

18

Burs use their tiny bristles to latch on to a deer's coat. Hang on tight—we're going for a ride!

Lending a Hand . . . or Paw or Wing

Have you ever come back from a hike with prickly cockleburs on your socks? What a pain it is to pick them out!

Lots of plants produce fruits that can hitch a ride. They get stuck in the fur and feathers of animals. As the animals move, they carry the fruits along. Eventually the fruits fall out and release the seeds they contain.

The tastiest trick for spreading seeds is making yummy fruit, like cherries, plums, avocados, walnuts, and papayas.

Mmm . . . are you getting hungry? All sorts of animals eat fruits. They may spit out the seeds or simply swallow them. Remember, though, that seeds are tough. Most seeds pass through animal digestive tracts unharmed. They end up in animal droppings. Complete with their own fertilizer, they're ready to grow!

▶ Like many birds, a macaw will eat the juicy part of a piece of fruit and then toss the seed.

The Bottom Line | **Flowering plants have evolved many different ways to spread their seeds.**

Acorns are the fruits of oak trees. They are hard nuts with little caps on top. Inside is a single seed—sometimes two. Acorns are an important food for many wild animals. But hardly anything loves acorns more than squirrels do!

▲ An eastern gray squirrel munches an acorn, which contains the seed of an oak tree.

Acorns are too heavy for the wind to blow around. Oaks trees need animals like squirrels to carry their seeds to new places. In the fall, squirrels collect acorns and bury them. The squirrels store acorns in the ground so they'll have food to eat all winter.

Eastern gray squirrels live in places where two kinds of oak trees, red and white, usually grow. Scientists observed the squirrels burying acorns. They discovered that the squirrels bury a lot more red acorns than white ones. When the scientists studied the two kinds, they discovered why. Red acorns are much less likely to rot underground.

Estimating

The scientists observed how many acorns the squirrels dug up and ate during winter. They estimated that the squirrels ate about 25 percent of the acorns they buried. That means for every four acorns a squirrel buries, three may grow into new oak trees!

white oak acorn

red oak acorn

▲ Acorns usually ripen and drop to the ground in autumn, about the time an oak tree's leaves turn color.

Interpreting Data

The scientists who studied acorn-burying squirrels gathered data all winter long. Data for one squirrel are shown in the table below.

Month	Acorns Buried	Acorns Dug Up
August	787	0
September	1,094	0
October	1,246	0
November	40	154
December	0	165
January	0	206
February	0	187
March	0	106

Your turn! Study the table and then answer these questions.

1. How many acorns did the squirrel bury during the study?

2. During what month did the squirrel dig up the most acorns?

3. Why might the squirrel have needed to eat more acorns that month than any other?

4. What percentage of the buried acorns survived being eaten by spring? How does that percentage compare with the scientists' estimate?

A Life Cycle

Do you eat your apples whole? Or do you slice them into pieces? Any way you eat them, you've probably seen apple seeds.

There they are, in the middle of all that sweet, juicy flesh. The dark brown seeds are small and very hard. Try crushing one by squeezing it. You simply can't.

If that tough little apple seed could talk, it wouldn't say much right now. It's **dormant**. That's a lot like being asleep. The embryo inside the seed is waiting. It's waiting for the right conditions so its **life cycle** can begin.

What starts the cycle? In the case of an apple seed, being planted in rich, moist soil can start the process. The seed soaks up water and swells. The seed coat splits, and **germination** begins.

▲ **How does an apple come to be? It starts with a seed.**

From Seed to Seedling

The embryo pushes out of its small seed home. Tender and pale, it begins to grow. Stem and leaves head up. Roots grow down. The energy to grow comes from the food stored inside the seed—the energy bar, remember?

Soon the embryo's stem peeks out of the soil. In the warm sunshine, its leaves turn green. The **seedling** starts to make its own food through photosynthesis.

The seedling grows into a tree. The tree gets bigger every year. Once spring comes, flowers form and open. The sweet scent of apple blossoms fills the air. So does the hum of honeybees. Bees visit the flowers in search of nectar. They transfer pollen from one flower to another.

▼ Bees can't resist the sweet-smelling flowers of apple trees.

The Wow !

Buzzing About

Honeybees really are busy. On an average day, a bee makes 10 to 15 trips between its hive and flowers, like the ones on a blooming apple tree. A bee may visit between 50 and 1,000 flowers on each trip!

The Bottom Line	The life cycle of an apple begins when an apple seed germinates and grows from a seedling into a tree.

Forming Fruit

What happens next is a really big deal. But you need a microscope to see it.

Pollen that the bees deposited on an apple flower's sticky stigma forms long, skinny tubes. The tubes grow down into the ovary, where the ovules are. Sperm from pollen move out of the tubes to join egg cells in the ovules. *Ta-da!* Special cells have just been formed that will become apple seeds.

The part of the flower that surrounds the developing seeds then begins to swell. It slowly changes into the fruit we call an apple. The apple starts out green. It gets bigger and bigger. When the apple reaches full size, it ripens. It gradually turns from green to red.

Deep inside the apple, the seeds are growing and ripening, too. They have changed from tiny pale specks into small oval shapes surrounded by a dark brown seed coat.

▶ **These photos show how the part of an apple blossom encasing the seeds gradually swells and ripens into a shiny red fruit.**

Round and Round

Soon the branches of the apple tree are heavy with fruit. As the apples get ripe, they fall to the ground.

Deer come to eat the apples. They bite and chew and swallow the fruit, seeds and all. Then the deer move on.

The next day, the deer's droppings are full of apple seeds. Some are nibbled by mice. Some rot. But a few get covered by soil. And then they wait.

The apple seeds wait as fall brings frost. They wait while winter brings cold and snow. Then the ground melts and warm spring rains arrive.

Hidden in the soil, the dormant seeds wake up. And the life cycle begins all over again.

▶ Deer love apples, and that's a good thing for apple trees. The deer spread apple seeds in their droppings.

So THAT's Why!

Why are flowering plants so important? They give us a lot more than just apples. Grains, beans, fruits, nuts, vegetables, herbs, spices, tea, chocolate—it's a really long list. Flowering plants put the cotton in your jeans. They give us medicines, dyes, inks, waxes, and building materials, too. Life without them would be tough!

The Bottom Line | An apple's life cycle is complete when new seeds form and fruit develops around them.

How Old Is the Orchid?

Do you suppose a dinosaur ever sniffed the sweet scent of an orchid? Since these lovely flowers are largely missing from the fossil record, it was hard for scientists to figure out whether orchids even existed back then. Some suggested that the orchid family could be as young as 26 million years old, while others said the plants might have been around much, much longer. Without fossils, it was difficult to say.

Enter the bee expert . . .
At first, Santiago Ramírez didn't realize how important the fossil was. He was looking at a bee fossilized in amber. The bee buzzed for the last time at least 15 million years ago.

Then it found itself in quite a sticky situation. The little bee got stuck in goopy tree sap, which covered it and eventually hardened into amber. Looking at the preserved bee was like looking back in time.

What made the bee really special was the cargo on its back—fossilized pollen grains. Working with an orchid expert, Santiago figured out that ancient orchids made the pollen. Amazing! Not only was this the first orchid fossil ever found, but it was also encased for all time with its pollinator!

▼ **Santiago examined the bee preserved in this ancient piece of amber, which was found in a mine in the Dominican Republic.**

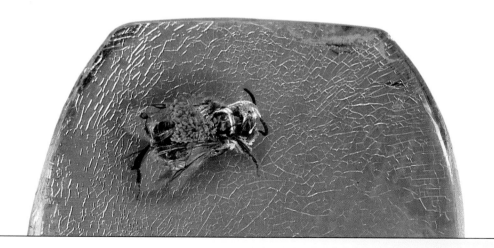

Santiago Ramírez

Evolutionary Biologist

UNIVERSITY
OF CALIFORNIA,
BERKELEY

▶ **Male orchid bees collect scent chemicals in little pockets on their hind legs and present them to female bees. How romantic!**

Santiago Ramírez grew up surrounded by animals on his family's farm in Colombia, South America. On a spectacular school field trip, he spent a week hiking and studying nature in the tropical forests near Colombia's Pacific coast. Something clicked. "I remember thinking, 'This is the kind of thing I want to do for a living,'" Santiago says.

In college, while looking for a research project, Santiago heard that an American researcher was planning a trip to the Amazon to study bees.

Santiago was invited to join the team. He has been studying bees ever since.

Today you might find Santiago in the lab tending to his colony of orchid bees. You might spot him hiking, skiing, or windsurfing in his free time. Or you might very well find him doing fieldwork in the forests of Central America.

◀ **Come right this way! To lure pollinators, the central petals of orchids are often the showiest.**

Like a detective, Santiago has been working with an orchid expert, gathering evidence. He has looked at differences in the DNA of various orchids and pieced together an orchid family tree. The branches show where orchids have evolved and formed new species.

Using a microscope, Santiago examines clumps of pollen on the back of the fossilized bee. "These grains are tightly packed," Santiago says. He points to a branch on the orchid family tree. "All the evidence suggests that our orchid belongs right here."

Based on the age of the amber, Santiago knows that the pollen and its branch of the tree are about 15 million years old. Since he knows how the rest of the branches differ genetically, he can estimate ages for the rest of the orchid tree. After piecing together all the evidence . . . *wow!* Santiago figures that the oldest orchids lived about *80 million years ago,* when dinosaurs still roamed Earth!

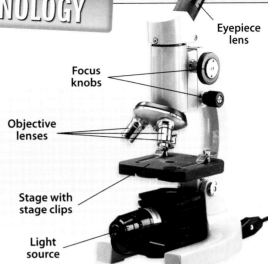

Eyepiece lens

Focus knobs

Objective lenses

Stage with stage clips

Light source

Santiago needed a closer look at the tiny fossilized pollen grains to see more details. The dot over this "i" is about 50 times larger than the pollen he was studying! Santiago used a compound microscope like this one to make the grains appear 400 times larger than they look to the naked eye. *Presto!* He could see them.

Sneaky Flower

Orchids have evolved unique ways to get insects to spread their pollen. This bucket orchid has a particularly bizarre trick. Part of its flower forms a little pool, which it fills with fluid. Bees are drawn to the orchid's scent and then slip into the pool. *Gotcha!* Their only escape route is through a side hole, where they pick up sticky pollen grains on their way out.

INVENTION CONNECTION

Dream Up a Flower
We all know orchids don't think when it comes to choosing pollinators. But imagine that you are an orchid. Come up with a clever way to attract a pollinator. How would you do it?

> First, imagine the insect or other animal you want to attract. Sketch a picture of your pollinator.

> Now, draw your flower. What shape would work best with your pollinator? What color would your petals be? What kind of scent would you give off? Why?

Hey, I Know That!

Okay, so you've cycled through this book at least once. You've met the flowering plants and seen what they can do. And you'll probably never look at an apple in quite the same way again! Now it's time to show what you know about these amazing plants. So grab a sheet of paper and do the activities and answer these questions.

1. What is photosynthesis? What was carrying out photosynthesis 500 million years ago? (page 6)

2. What are the parts of a seed? Why is each important? Why was seed-making such an important development for ancient plants? (pages 10 and 12)

3. Copy the chart onto a piece of paper. Match the type of plant with its characteristics. Write the matching letter under the appropriate number. Then pick one type of plant from the chart. Draw what one looks like and write three sentences about it. (pages 7–13)

Type of Plant			Plant Characteristics
1	Cone-bearing	A	• flat, low-growing • reproduces with spores • lives near water's edge
2	Flowering	B	• needs water to reproduce • helped to form first forests • found with giant horsetails
3	Tree ferns	C	• reproduces with spores • small and branching • lives only in moist places
4	First land plants	D	• can live almost anywhere • makes seeds and fruits • need pollinators
5	First plants with vessels	E	• reproduces with seeds • produces pollen • seeds sheltered by cones

4. What is your favorite flower? Draw and label its parts. Where are the ovules located? (page 15)

5. Give an example of a pollinator and explain how it helps a flowering plant reproduce. (page 16)

6. Name at least two adaptations that flowering plants have that attract pollinators to their flowers. (page 19)

Glossary

adaptation (n.) the evolution of features (anatomical, physiological, or behavioral) that make a group of organisms better suited to live and reproduce in their environment (p. 16)

alga *singular* **algae** *plural* **(n.)** a usually single-celled eukaryotic organism that carries out photosynthesis. It is a member of protist kingdom. (p. 6)

cone (n.) the reproductive structure of conifers—cone-producing trees with needle-like or scale-like leaves (p. 12)

dormant (adj.) in a resting state (p. 22)

evolution (n.) changes in the collection of genetic material, or the gene pool, from one generation to the next as a consequence of natural selection and other processes (p. 7)

fertilization (n.) the coming together of a male sperm cell and a female egg cell to form a fertilized egg (p. 11)

flowering plant (n.) a plant that has flowers and reproduces by making seeds protected inside fruits (p. 5)

fossil (n.) the remains or traces of an animal or plant preserved in some way, usually in rocks, but also in ice, peat, or tar (p. 9)

germination (n.) when a seed splits open and the embryo inside begins to grow (p. 22)

habitat (n.) a place where individual organisms of a particular species live. Each habitat has a particular set of conditions and array of other species. (p. 14)

life cycle (n.) the entire span of existence of a living organism (p. 22)

nectar (n.) a sugary liquid produced by flowers that attracts pollinators (p. 16)

nutrient (n.) a substance that plants and animals need to grow and survive, usually obtained from food (p. 18)

ovule (n.) the female reproductive structure in flowering plants in which egg cells form. After fertilization, it matures into a seed. (p. 11)

ozone layer (n.) a layer of ozone (O_3) molecules in the stratosphere that blocks some ultraviolet rays coming from the Sun (p. 7)

photosynthesis (n.) the process by which plants and other photosynthetic organisms use energy from sunlight to make sugar from carbon dioxide and water. As part of this process, oxygen is released. (p. 6)

pollen (n.) a male reproductive structure made up of a sperm cell and a protective outer covering. It develops within the anthers of a flower. (p. 11)

pollinator (n.) an animal that helps flowering plants reproduce by transporting their pollen from flower to flower (p. 16)

reproduce (v.) to make more of one's kind (p. 6)

seed (n.) a structure made by plants that contains the embryo of a new plant (p. 10)

seed coat (n.) the tough protective covering of a seed (p. 10)

seedling (n.) a very young plant that has recently come out of the seed (p. 23)

sperm (n.) a male reproductive cell (p. 11)

spore (n.) an asexual reproductive cell that can develop into an adult plant (p. 6)

ultraviolet light (n.) a region of the electromagnetic spectrum with wavelengths shorter than violet light and longer than X-rays. Ultraviolet light is energetic enough to damage cells. (p. 7)

vessel (n.) a tube-like structure made up of cells that carry food and water inside a plant (p. 8)

Index

About the Author Rebecca L. Johnson is a national award-winning author of more than 70 books for children and young adults about science. To learn more visit www.SallyRideScience.com.

Photo Credits Terry Youngblood: Cover. Valentyn Volkov: Back cover. Michael Quinton/Minden Pictures: Title page. Holly Kuchera: p. 2. Paul Maguire: p. 5 (flower). Carolina Biological Supply, Co./Visuals Unlimited: p. 6. Todd Boland: p. 7. Dr. Jeremy Burgess/Science Photo Library: p. 8 bottom. © 1995 Saint Mary's College of California: p. 9 bottom. Steve McWilliam: p. 11 top. Enrico Boscariol: p. 11 bottom. Cheryl Wuschke: p. 12 bottom. Marketa Mark: p. 14. Frans Lanting: p. 15. © Merlin D. Tuttle, Bat Conservation International: p. 16. Willi Schmitz: p. 17. Blaz Kure: p. 18 (maple seeds). Brian A. Jackson: p. 18 (dandelion). Laure Neish: p. 19 (deer). Thomas Marent/Minden Pictures: p. 19 bottom. Joe Gough: p. 20 (squirrel). Steve Hurst, USDA-NRCS Plant Database: p. 20 (white oak acorn). Tomasz Kopalski: p. 20 (red oak acorn). Amy Walters: p. 22 (bushels of apples). Mark Longstroth: p. 24 top and middle. Frank Leung: p. 25 left. Varga Levente: p. 25 right. Courtesy of Santiago Ramírez: p. 26, p. 27 (bee), p. 28 (Santiago at desk). J. Bermudez: p. 27 (Santiago hiking). Courtesy of University of California, Berkeley: p. 27 (logo). Lijuan Guo: p. 28 (orchid). Microscope World: p. 29 top. Raymond Prothero Jr.: p. 29 bottom.